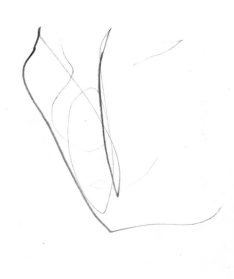

Seymour Simon

SEE MORE READERS

FIGHTING FIRES

SeaStar Books ● New York

This book is dedicated to Robert and Nicole.

Special thanks to reading consultant Dr. Linda B. Gambrell, Director, School of Education, Clemson University. Dr. Gambrell has served as President of the National Reading Conference and Board Member of the International Reading Association.

Permission to use the following photographs is gratefully acknowledged:
front cover, pages 6–7: © Mark L. Stephenson/CORBIS; title page: © Peter Skinner, Photo Researchers, Inc.; pages 2–3: © Gary Braasch/CORBIS; pages 4–5: © Paul A. Souders/CORBIS; pages 8–9, 12–23, 26–27, 32: © George Hall/CORBIS; pages 10–11: © Gunter Marx Photography/CORBIS; pages 24–25: © David Rowley, New England Stock; pages 28–29: © Michael S. Yamashita/CORBIS; pages 30–31: © Peter B. Kaplan, Photo Researchers, Inc.

Library of Congress Cataloging-in-Publication Data is available.
ISBN 1-58717-168-6 (reinforced trade edition)
1 3 5 7 9 RTE 10 8 6 4 2
ISBN 1-58717-169-4 (paperback edition)
3 5 7 9 PB 10 8 6 4 2
PRINTED IN BELGIUM
For more information about our books, and the authors and artists who create them,
visit our web site: www.northsouth.com

Firefighters risk their lives to put out fires and save other people's lives.

Long ago, people passed buckets of water by hand to put out fires. Now, when an alarm sounds, firefighters have many kinds of special trucks to help them.

Pumper trucks carry water, hoses, and a pump. In just one minute, pumpers can shoot out 1,250 gallons of water.

That's enough water
to fill 50 bathtubs.
Many pumper trucks
also make bubbly foam
that puts out oil fires.

Some pumper trucks carry
a long crane called a boom.
Firefighters stand inside
a large bucket and are raised
100 feet into the air
to pump water or foam
on a fire.

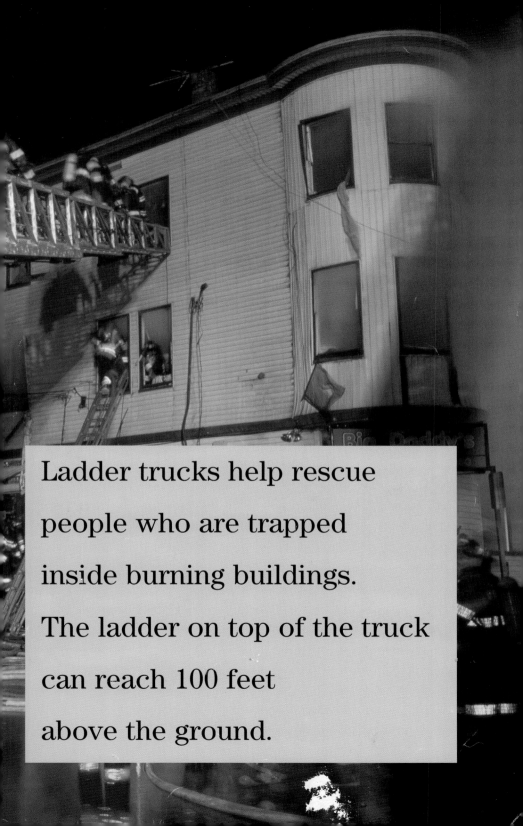

Ladder trucks help rescue people who are trapped inside burning buildings. The ladder on top of the truck can reach 100 feet above the ground.

Tiller trucks are two ladder trucks that are joined together. One driver steers the front and a tiller driver steers the back.

Two drivers make it easier for the big truck to get through narrow city streets.

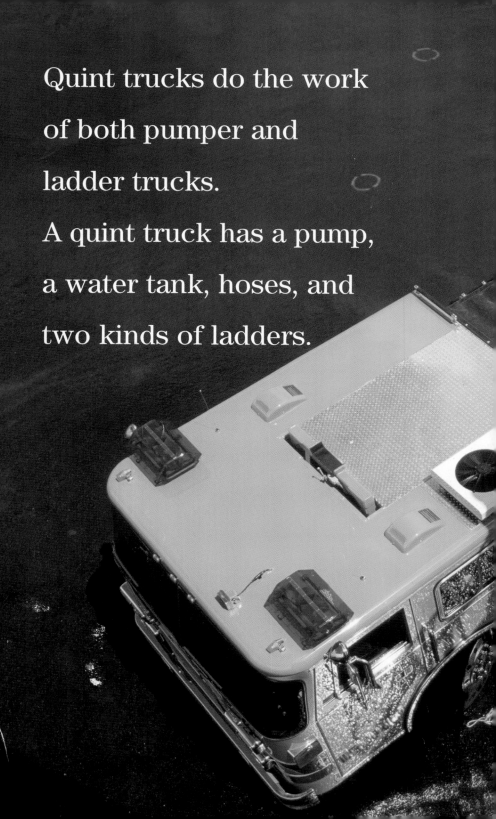

Quint trucks do the work
of both pumper and
ladder trucks.
A quint truck has a pump,
a water tank, hoses, and
two kinds of ladders.

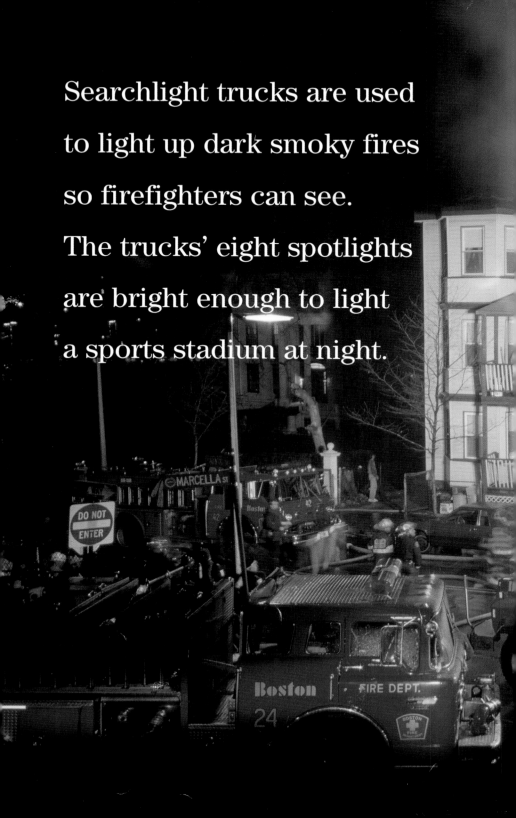

Searchlight trucks are used to light up dark smoky fires so firefighters can see. The trucks' eight spotlights are bright enough to light a sports stadium at night.

Rescue trucks carry tools
for emergencies, including
special suits for walking
through fire.

Rescue trucks also use the "Jaws of Life" to help remove people from crushed cars or collapsed buildings.

Aircraft crash rescue vehicles carry as much as 2,500 gallons of water.

That's because an airplane crash may be a long way from a water supply. A crash rescue vehicle can put out an airplane fire in minutes.

Off-road fire trucks

are used to fight forest fires

and brush fires.

They carry water, foam,

hoses, and pumps.

Off-road fire trucks need to be tough enough to drive in places where there are no roads.

Special helicopters drop water
from 100-gallon buckets
onto forest fires.
A huge bucket scoops up water
from a lake for the next flight
over the flames.

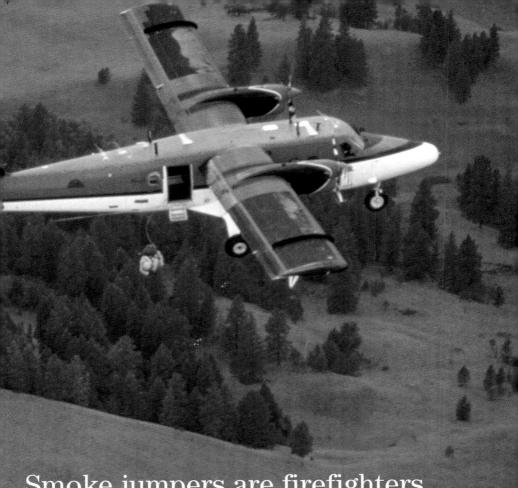

Smoke jumpers are firefighters who parachute from airplanes. Their work is very dangerous. They have no way to escape if the wind shifts and the wildfire burns toward them.

Fireboats put out fires on ships or in waterfront buildings. A fireboat does not carry water, it takes it from beneath the boat. A big fireboat can pump as much water as ten pumper trucks.

Special trucks, boats, and aircraft help us fight fires.